Creation Tithing
a Primer

Anne Ipsen and Ken Kroes

CREATION TITHING, A PRIMER

ISBN 978-0-9828778-7-6
First Edition: July 2018

Cover design by Ken Kroes.

ibus press, Cambridge Massachusetts
www.CreationTithe.com

How to Order

Information on ordering multiple copies of this primer, at a reduced price, can be found on www.CreationTithe.com. The website can also be used to download support materials, including copies of the tracking tables, which may be reproduced or adapted without special permission.

Table of Contents

Introduction

"So God did just that. [God] created humanity in [God's] image, created them male and female. Then God blessed them and gave them this directive: "Be fruitful and multiply. Populate the earth. I make you trustees of My estate, so care for My creation and rule over the fish of the sea, the birds of the sky, and every creature that roams across the earth." Genesis[1]

What is Creation Tithing?

Creation Tithing is a program for houses of worship and their members to reduce greenhouse gas emissions by 10% each year through a dedicated focus on reducing the worst consumer-generated emitters: electricity, gas, heating oil, water, and automobile fuel.

Terminology

We use the word *participant* to describe an individual, family, or other groups that participate in the creation tithing program as a unit or *household*. The house of worship of a sponsoring faith community is abbreviated *HOW*. The HOW includes the parish hall, parsonage, and any other buildings that it owns. *Clergy* stands for worship leaders, *Council* for the vestry, trustees, or committee that administers the community and its property, and *congregation* for the members of the HOW.

The five household *utilities* named in the definition of Creation Tithing: electricity, gas (natural or propane), oil, water, and automobile fuel (gasoline or diesel), emit greenhouse gas in the form of *carbon-dioxide* (CO_2). A participant may not necessarily use all five utilities but will *track* the ones they do from their bills. Usage will be measured, not in terms of dollars, but according to the units used on the bill, be that in kilowatts, thermal units, or gallons. The *appliances* in our household (or the HOW) are the actual emitters—yes, even a shower will be called an appliance, as will the heating system and your automobile. In the first year (the *baseline*), the *goal* for each utility is set to 10% less than it was the year before. Each subsequent year the new goal is an additional 10% less. For how to set your goal see Chapter 3.

How to order the Primer

Information on ordering multiple copies of this primer at a reduced price can be found on www.creation.tithe.com. The website can also be used to download support materials, including copies of the tracking tables, which may be reproduced or adapted without special permission.

How to use the Primer

A Creation Tithing program is designed to be a community effort that involves the HOW, its clergy, and members of the congregation. It is to be a **sustained effort** in which the participants plan their own specific goals and support one another with periodic get-togethers. Just as the most successful weight-loss programs are designed around a systematic plan and mutual support, our carbon-reducing program depends on doing it together.

A weight-loss program may consist of a food plan with a short-term goal of a specific weight loss, tracked by means of a food diary, and holding the participant accountable on the scale. In Creation Tithing, we *plan* how to use our appliances, our *goal* is our tithe, and we *track* our utility bills and use as the scale by which we hold ourselves accountable. Just as the long-term goal of a good weight-loss program is meant to instill permanent change towards better nutrition, the long-term goal of Creation Tithing is to change our consumer habits to a lifestyle that supports and sustains creation for the future.

> **Hindu Prayer**
> My salutations to you, O Bhudevi [Mother Earth], consort of the all-pervasive Lord, forgive me for placing my feet upon you.

All major religions agree that we must be good stewards of our planet. While the language of the primer is primarily targeted for use by churches, the intent is to include all faith communities, as well as individuals, families, or informal groups. The tithing program can be sponsored by a single faith community, or by cooperation among several, and the public may be invited to participate. We particularly encourage families to include their children and youth as much as is practical—they have the biggest stake in the future and may turn out to be the most enthusiastic participants.

The primer is a first step, and a tool for generating ideas through discussion. It can be used by workshop leaders to develop their own materials, as a framework for workshop slides, and as a workbook for individual participants. It is intentionally brief and does not pretend to be comprehensive nor definitive. The end of the book (Appendix II) contains a bibliography of sources for more extensive discussion of the topics introduced, and the close of each chapter cites the quotations and websites referenced. While there are many websites[2] and books[3] regarding reducing carbon footprints, this primer is slightly different. It provides resources for supporting Creation Tithing through worship, community-building, connections to the land, and justice activities, keeping the original intentions of tithing intact. As people of faith, we undertake Creation Tithing not just as a duty, but through a passion to deepen our love for God, live out our call to care for our Created home, and "to do justice, and to love kindness, and to walk humbly with [our] God."[4]

End Notes for the Introduction

[1] Bible (The Voice translation) Genesis 1:27-28, "He" replaced with "[God]"

[2] **Websites**: http://www.interfaithpowerandlight.org/about/state/, http://www.mipandl.org, https://www.macucc.org/carbonfast

[3] **Books** (complete citations in Appendix II): "Cooler Smarter: practical steps for low-carbon living", "Feasible Planet, a Guide to more Sustainable Living", "Low Carbon Diet: a 30-day program to lose 5,000 pounds"

[4] Bible (NRSV translation) Micah 6:8, "your" replaced with "[our]"

Chapter 1. WHY CREATION TITHE

All tithes from the land, whether the seed from the ground or the fruit from the tree, are the Lord's; they are holy to the Lord. Leviticus.[5]

If you are reading this primer you will not need to be convinced that global warming is a problem caused by modern civilization, but you may not realize that climate change **is already** a crisis that affects us today, not just an issue that will touch our children and grandchildren in the future. In 2017 alone, the triple hurricane catastrophes in the Caribbean Islands and Gulf coast states, the monstrous fires in California, the melting of arctic ice, floods, droughts, and famines around the globe are all dire warnings of a situation that will only become worse.[6] We have squandered the bounty of God's Creation by carelessly using and abusing its resources without regard for the consequences.

Our environmental leaders tell us that we must emit only 720 gigatons more CO_2 during this century to avoid exceeding a global rise in temperature of 2ºC.[7] At the current rate of emission this "carbon budget" will be used up by 2035, within most of our lifetimes.[8] To live in hope for a viable future, we must make a total commitment now to "change everything" in our overconsuming lifestyles—all the ways we live, work, and interact with Earth.[9]

Our religious leaders tell us that we are facing the biggest moral challenge and opportunity of the twenty-first century.[10] Despite the economic and social challenges currently facing them, congregations and their leaders have access to spiritual practices and have traditions of prophetic witness. Our faith communities can become active agents of change for climate justice and lead the way in this moral, spiritual, environmental, and political crisis.

It is soul-shaking for each of us to face up to our human responsibility for causing this mess, especially in North America. We are called to deeply repent of the harm we have done to Creation (called "good" by God) and the harm this has done to our relationship with the Creator. It would be easy to become paralyzed by guilt, overwhelmed by the complexity of the problem, and protest the busyness of our lives. We despair and cling to the hope that our governments and businesses will change and act responsibly. But we are people of faith, which means that we must hope, pray, AND ACT.

This primer describes a program for committing ourselves to change by focusing on the worst carbon emitters under our consumer control: the utilities that run and heat our households and houses of worship, and the fuel that powers our cars. Since we pay directly for the electricity, gas, heating oil, water, and gasoline that we consume, we can set specific goals for these utilities and thus hold ourselves accountable for reducing them and our carbon footprints by ten percent. Everyone is on an equal footing with no one allowed to rest on their laurels. It does not matter what you have done or not done in the past as the goalpost moves each year and we each set new reduction goals.

Of course, other consumer goods also have a huge carbon footprint, but we cannot as easily set goals nor track our usage. Neither can we, nor should we, ignore the effect we can have on public climate policy through the activist organizations we join, the politicians we support and for whom we vote. These other aspects of Creation Care will be briefly discussed in Chapter 4, but to keep it simple, the remaining chapters will focus on tithing the utilities named above.

> **Other Environmental Culprits:**
> Food, clothing, transportation,
> AND
> the Public Policies that regulate them

Origins of Tithing

The spiritual practice of giving back from the gifts we have been given by our Creator has a long history in many faith communities.

In Jewish tradition, YHVH/God commanded the Israelites to give a tithe (a tenth) of their crops, fruits, and herds each year to support the Levites (priest) and an additional tithe every third year to support the poor.[11, 12, 13] Some important points about Mosaic Tithing, as it is called, are:

> And [Abram] gave [the priest of God Most High] a tenth of everything.
> Genesis 14:20

- *It connected the Israelites to YHVH. Tithing was an act of worship, of strengthening the Israelites' relationship to their Creator and Sustainer.*
- *It connected the Israelites to the land. The land was YHVH's promise to the people, and the bounty of the land was YHVH's gift to the people. Food is necessary to life but is not something that humans can make on their own. Giving back from the yearly produce was a constant reminder of their dependence on God for life.*
- *It connected the people of Israel to each other. All farmers and herders of the community participated in tithing; it was what you did as an Israelite.*
- *It was an act of justice. Of all the tribes, only the Levites had no land; therefore, they received the tithe from the other tribes to support their work on behalf of the people.*

- *The additional tri-annual tithe provided for the poor in the community—resident foreigners, widows and orphans--providing sustenance to those who would otherwise go hungry.*

From earliest times, the principles of Biblical tithing were adopted by Christians to secure the welfare of the church community. The Roman Catholic Church required everyone to contribute to its support by donations of land or produce, sometimes in the form of a tithe. In addition to being houses of worship, the church operated hospitals, offered hospitality to travelers, and gave charity to those in need. In modern times, the state has assumed some of these social services paid through compulsory income taxes. Still, members of a church or synagogue contribute to its work in the community and to maintenance of the house of worship with an annual pledge. In some denominations, the pledge is a tithe or 10% of income, in others that is merely encouraged. It is expected that members also donate to organizations that help the poor or work on behalf of those in need.

Similarly, financially stable Muslims give an annual contribution of 2.5% of their wealth and assets. This contribution, called *Zakat*, is commanded by *Allah*, as mentioned in the Qur'an, and is considered one of the Five Pillars of Islam.[14,15] *Zakat* is an act of worship that reminds Muslims that everything they have belongs to God. In Arabic, the word literally means "purification and increase" because it is meant to purify one's heart of greed and increase the wealth of the community. In a separate form of Islamic charity, *Sadaqa*, every member of the community—rich and poor—is required to *give up* a blessing from *Allah*. This can consist of: time, knowledge, health, or effort.

Five Pillars of Islam

1. Testimony (*Shahada*)
2. Prayer (*Salat*)
3. Charity (*Zakat*)
4. Fasting (*Siyam*)
5. Pilgrimage (*Hajj*)

Creation Tithing is therefore similar to *Sadaqa* in that we give *up* instead of *back* in a voluntary tax on consumption rather than on income. Like Judeo-Christian Tithing and Islamic *Zakat*, it serves to strengthen our relationship to God, to the land we live on, the Earth we inhabit, our communal life together in faith, and our commitment to the just distribution of resources. Creation Tithing is not a temporary fast of doing without for the day of Yom Kippur, the month of Ramadan, or six weeks of Lent, but a weight-loss program for the planet that is offered in a joyful act of worship, as we strive for a healthful and permanent change in our lives away from greedy and thoughtless over-consumption.

End Notes for Chapter 1

---

[5] **The Torah,** Leviticus 27:30, The Jewish Publication Society of America, 1963.

[6] **Climate Science Special Report,** Fourth National Climate Assessment, (USGCRB), Wuebbles, DJ, DW Fahey, KA Hibbard, DJ Dokken, BC Stewart, and TK Maycock (eds). US Global Change Program, doi: 10.7930/JOJ964J6. Especially the shorter Executive Summary found at (https://science2017.globalchange.gov). (2017)

[7] ***Do the Math,*** *global warming's terrifying new math* by Bill McKibben, in *Rolling Stone* (http://www.rollingstone.com/politics/news/global-warmings-terrifying-new-math-20120719). (2012)

[8] https://www.mcc-berlin.net/en/research/co2-budget.html (as of 2/2017)

[9] ***This Changes Everything,*** *Capitalism vs. the Climate* by Naomi Klein, Simon & Shuster (2014)

[10] ***Our Moral Opportunity on Climate Change*** *by* Bishop Justin Welby, Ny Times op ed, 11/3/17, (https://www.nytimes.com/2017/11/03/opinion/faith-climate-change-justin-welby.html). Welby is the Archbishop of Canterbury and spiritual leader of the Anglican Communion. (2017)

[11] ***The Torah,*** *Genesis* 14:20, The Jewish Publication Society of America, 1963.

[12] ***The Torah,*** *Deuteronomy* 14:22-23, 28-29, The Jewish Publication Society of America, 1963.

[13] ***The Torah,*** *Numbers* 18:21-24, The Jewish Publication Society of America, 1963.

[14] **Qur'an** 7:156 and 21:73

[15] **The Pillars of Islam:** http://www.quranislam.org/main_topics/islam/pillars_(P1188).html

Chapter 2. Appliances and How We Use Them

I arise today, through the strength of Heaven; light of Sun, brilliance of Moon, splendor of Fire, speed of Lightning, swiftness of Wind, depth of Sea, stability of Earth, firmness of Rock...; through confession of the Oneness of the Creator of Creation.

The Breastplate of St. Patrick[16]

This chapter is an overview of each of the common household appliances, including heating and cooling systems. Together with our automobiles, they are responsible for most of the carbon emissions of a household or HOW, as shown in the table. Web-links to sources of information are at the end of the chapter and suggestions for more reading are at the end of the book. The website[17] of the Smarter-House organization (https://SmarterHouse.org) is full of useful hints, some of which have been inserted below. This organization is a project of the non-profit American Council for an Energy-Efficient Economy.

Home Energy Breakdown
https://SmarterHouse.org [17]

- *Heating - 26%*
- *Cooling - 17%*
- *Water Heating - 13%*
- *Appliances - 14%*
- *Lighting - 10%*
- *Electronic Devices (TV, Computers, etc.) - 7%*
- *Misc. Other - 13%*

Before you start setting up your tithing plan, sign up for a free or low-cost energy audit and seriously consider all their suggestions. In the U.S., HOWs can contact the non-profit Interfaith Power and Light (IPL) organization[18] by visiting their website (www.interfaithpowerandlight.org) and searching for the state affiliate. Households may be eligible for a free audit through their local energy supplier, e.g. MassSave in Massachusetts.[19] In purchasing new appliances, look for the Energy Star label. The manufacturers of these appliances participate in an EPA sponsored program that certifies those with high efficiency.[20]

Most Efficient 2018 ENERGY STAR www.energystar.gov

Each section below contains two kinds of suggestions: Low Hanging Fruit and High Impact Changes. We suggest that you go over the items in these lists and: check mark (√) those you have already done, cross out (or x) those that do not apply to you or are not feasible, and star (*) those that you will adopt or attempt next. There is space for notes in the margins and after each list.

The **Low Hanging Fruit** checklist consists of quick fixes that have a quick return, are low-cost, or free. Some may seem trivial, especially if you are more experienced in reducing your carbon foot print and have already carried out or considered most of these suggestions. Remember that even small additional changes can raise your consciousness and bear witness to others.

The **High Impact** suggestions consist of more advanced approaches that will have a large impact on carbon emissions but may involve more work and a higher cost with slower payback. Since a considerable financial investment may be necessary, explore if low or free-interest loans for energy improvements are available to your household or the HOW. These lists are just suggestions for where to begin and are not complete nor necessarily the "best." They are intended to stimulate discussion, raise questions, and generate ideas. A few references for further reading can be found in Appendix II at the end of the Primer.

Heating, Ventilating, and Air Conditioning (HVAC) System

New or recently renovated houses may have a combined system for heating and cooling, called an HVAC (heating, ventilation, and air conditioning). In older homes the different functions may be done by separate units. In a typical household, the HVAC system may account for almost half of the energy use, and in your HOW, it may be an even larger portion of the energy (and fiscal) budget. Replacing the HVAC with a high efficiency system may be the best help you can personally do for creation and a sound long-term investment.

The heating part of the system may consist of a furnace or boiler that heats water or air that is distributed by a pump or fan throughout the home or HOW. The system may be powered by electricity, natural gas, propane gas, or oil. Other sources, with higher global-warming emissions, are wood and coal. We call this ordered list of increasing emission *the emission-ladder*. The heat may be distributed throughout the house by hot water radiators or an air ventilation system. Radiators are generally more efficient and provide a more consistent temperature but are more expensive to install, whereas hot air systems are easier to combine with central air conditioning.

Air conditioning (A/C) comes in two varieties: individual units or a central system. Individual units consist of a condenser and a fan and are primarily used to cool only one room. Each unit is either inserted in a window or installed in an opening in the wall of that room. A central system consists of one or more condensing units, depending on the number of cooling zones, installed outside the building. A ventilation system pumps the air around indoors and may be part of the heating system. Both kinds of A/C types are controlled by thermostats, one for each condenser, and are powered by electricity. In houses with radiators for heating, there may be a separate central A/C system.

The temperature for both heating and cooling is usually controlled with an electrical thermostat, which may be programmed for settings for different days of the week and separate settings for heating and cooling. The space may be divided into separate zones, each with its own thermostat.

> A new HVAC installation can include "whole house" fans that bring outside air indoors. The ventilation system pumps cooler night air (in the summer) or, (in fall and spring) warmer daytime air into the house, thus reducing the use on heating or cooling.

The most energy efficient but costly to install are **heat exchange systems** that combine heating and cooling functions. They too can be small units meant for single rooms. For new construction, consider a geo-thermal system that uses water that has been cooled (in summer) or pre-heated (in winter) in the consistent temperature deep underground.

HVAC, whether it consists of separate appliances or a comprehensive system, represents a considerable investment but is responsible for the major portion of a household's carbon emissions. Replacing the system or individual units with newer models with higher efficiency ratings, preferably an Energy Star rating, may save money in the long run. Also consider the time to payback and explore state supported low or zero-interest energy loans.

Low Hanging Fruit—HVAC system

- ☐ Maintain the system(s) as recommended by the manufacturer for peak efficiency and longer life:
  - Replace air filters regularly.
  - Schedule regular maintenance by a professional.
- ☐ Install either a programmable or smart thermostat. Programmable thermostats allow you to set a schedule for your home temperature. Smart thermostats are easier to program and "learn" your habits for setting temperature and will soon start to set it automatically.
- ☐ Set thermostat one degree below (winter) or above (summer) your comfort level for when you are home and three to five degrees for periods of the day when you are regularly absent.
- ☐ When you are traveling and will be gone for more than a day, set thermostat on "hold" to maintain a constant temperature 5-10º outside the usual setting.
- ☐ Close air vents, radiator valves, or turn off window A/C units in rooms that are seldom used and close the door.
- ☐ Weather-strip windows with tape or spray with a special sealant.
- ☐ Add "draft-dodgers" to base of doors to the outdoors.

- ☐ Insulate electrical sockets on outside walls.
- ☐ Help control temperature with window shades.

  In summer: lower shades on south windows during the day.

  In winter: lower all shades at night, raise on south windows during sunny days.
- ☐ Plant deciduous shade trees on the south side of house.
- ☐ Keep storm windows in place year-around.

Low Hanging Fruit—additional for A/C

- ☐ If the outside temperature is lower than that indoors, open the windows in several parts of the house and switch the fan from "auto" to "on".
- ☐ In very hot weather ,when the A/C is on, switch the fan from "auto" to "on" to keep the air circulating.
- ☐ Consider turning off the A/C except in extremely hot weather.

Higher Impact—HVAC

The most effective way to lighten your carbon footprint is to upgrade an outdated HVAC system in one or more of the following ways:

- ☐ change your heating system so that the power source moves down the *emission-ladder* one or more steps, i.e.: coal => wood => oil => propane gas => natural gas => electricity. Be aware that the carbon footprint of electricity depends on how it is generated by the power company, see the section on **Electricity** in Chapter 3.
- ☐ install boilers, condensers, and/or A/C units with efficiency ratings of 90% or more.

- ☐ add zones for different parts of the building.
- ☐ larger buildings, such as a HOW, may benefit from a dual boiler system.
- ☐ consider installing European-type radiators that have individual thermostats.
- ☐ consider installing radiant heat.

Improve the insulation (likely to be cost effective with quick payback):

- ☐ add more insulation to the attic or the building envelope.
- ☐ replace drafty windows with double or triple glazed panes.
- ☐ add weather stripping or storm windows to each window opening.

Additional considerations:

- ☐ Even if the A/C system is central, install window units in bedrooms for separate cooling at night.
- ☐ Install an attic fan that will draw cool night air into the house during the summer.
- ☐ When building a new home or extensively renovating an older one, use a building contractor that specializes in energy efficient building materials. [21]

Hot Water Heater

The system typically consists of a 40-60-gallon tank, heated by either gas or electricity, with the temperature controlled by a thermostat. If the HVAC system uses hot water, the same tank may also provide hot water for use by appliances. The typical life expectancy of a tank is 8-10 years and the cost is around a thousand dollars to replace. It may account for 13% or more of the household's consumption of energy [see Table in Display 1.1]. Consider replacing with a heater that is more energy efficient and better insulated, even if the old unit is working, and install as small a tank as is practical. For HOWs, which typically do not have showers (except in associated housing for clergy), a 15-20-gallon tank may be sufficient for washing hands and dishes. Most of the other fixes are low-hanging fruit, but how we use the *appliances* that need hot water is important; see the next section.

Low Hanging Fruit—hot water heater

- ☐ Set the thermostat to heat water to 120ºF, the lowest temperature that prevents mold from growing (if you have a suppressed immune system or chronic respiratory disease, set it to 140 ºF).
- ☐ Reset the temperature to "vacation" or lowest setting when living elsewhere for more than a few days.
- ☐ Consider installing additional very small, "on-demand" heaters in sinks, especially those far from the main hot water heater or where only small amounts of hot water are needed and only rarely. This may be especially relevant for a HOW with remote washrooms.
- ☐ Manufacturers recommend regular professional maintenance, including draining the tank one or more times a year to prevent calcium buildup and extend life.

Household Appliances that use Hot or Cold Water

In the U.S., the average daily consumption of water is 80-100 gallons per person,[22] mostly from use of household appliances and watering the lawn and garden. Although the energy required for processing and delivering water and treating sewage does not contribute significantly to green-house gas emissions, water is an increasingly scarce resource, and worthy of our attention and tithing (see the section on Water in Chapter 3).

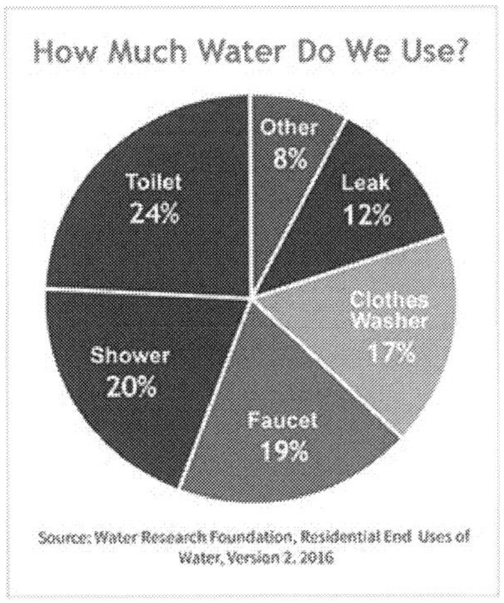

Flushing the *toilet* and *watering* the lawn contribute the largest share of water usage, but your *shower, sink faucets, dishwasher,* and *washing machine,* use *hot* water.[23] The last two are also powered by electricity and replacement with ones that have a high efficiency or EnergyStar rating and using them sparingly will cut down significantly on both water and energy consumption.

The following appliances consume the most water: *toilets, dishwasher, washer/dryer, shower, sink faucets,* and *watering outdoors*. Their carbon-footprints depend very much on how long and how frequently they are used and whether or not they require both water and electricity to operate.

Low Hanging Fruit—appliances that use water

- ☐ Stop the leaks! Ensure that all faucets are leak-free. Check each toilet by putting a drop or two of food coloring in the tank and check in a few minutes to see if the water in the bowl changes color.
- ☐ **Toilets** are the largest consumers of water because of their frequent use. Each flush requires 1.3 to 4 gallons of water or more, depending on the age and style of the toilet. Consider flushing only when waste is solid. Or try the following trick: when waiting for the water to get hot in a sink or shower, collect the water in a pitcher or bucket. Flush toilet by *dumping enough water in the bowl to start the flushing process*. If you have an older, high volume model of toilet, consider replacing it with a dual flush model. The small flush on these can be as little as 0.8 gallons. Another alternative is to put a full water bottle or brick in the tank which reduces the amount of water used per flush.
- ☐ **Dishwasher**. Turn off the "heated dry" option and run only when full. Do not scrub the dishes before putting them in the dishwasher but scrape and rinse minimally. If you have only a few dishes, wash by hand and air-dry in dishwasher rack.
- ☐ **Washing Machine**. Run only full loads or schedule regular days for accumulated light-colored vs. dark-colored laundry. Wash only in warm or even cold water and spin dry at highest setting. For special small loads, reduce the water level (if possible).
- ☐ **Dryer**. Reduce dryer use by only drying clothes for a few minutes then hanging them to air dry. Or hang directly from the washer, outside (if possible), otherwise on a drying rack (placed in the bathtub) or on hangers (on shower-curtain rod).
- ☐ **Shower.** Shower only 2-3 times per week and only for 5-10 minutes. For showers in bathtubs, you can avoid having to use a timer by leaving the plug in during a test run and notice the height of the accumulated water. Remember to reserve the initially cold water for flushing the toilet or watering house plants. Another idea is to

replace the shower head with a low flow one which can save a ½ gallon or more per minute of shower time.

- ☐ **Sink Faucets**. Turn off the faucet when not needed and use cold water whenever possible. When hot or warm water is necessary, collect the initially cold water in a pitcher for drinking or watering house plants. Also consider replacing the filter on the tip of the faucet with low flow variety.
- ☐ **Outdoor Watering**. Avoid watering during the day and only when needed. Place mulch around your shrubs. During drought conditions, water lawns sparingly—brown grass will come back. Consider minimizing grassy areas by landscaping with more drought resistant plants. This will also reduce the need for high-emission mowing.

Household Appliances used for Food Preparation

Most households have a range or stove for both cooking and baking. It may be powered by gas (natural or propane) or by electricity, or a combination. In addition, many households have a microwave, toaster, toaster-oven, hot water kettle, and/or slow-cooker. These are all powered by electricity and are generally more efficient than their larger counterparts.[24]

Energy Used by Cooking Appliances
SmarterHouse.org [19]

APPLIANCE	TEMP.	TIME	ENERGY
Electric Oven	350ºF	60min	2.0 kWh
Convection Oven	325ºF	45min	1.4 kWh
Gas Oven	350ºF	60min	0.11 therms
Frying Pan	420ºF	60min.	0.9 kWh
Toaster Oven	425ºF	50min	0.95 kWh
Slow Cooker	200ºF	7hrs	0.7 kWh
Microwave Oven	"High"	15min	0.35 kWh

A comparison of heating one mug of water shows that an electric kettle is almost three times as efficient as using the stove and nearly twice as efficient as the microwave. However, when heating water for beverages in the microwave, the mug is also heated (keeping the drink warm longer) and it is easier to heat only the quantity actually needed. Using a pressure cooker reduces cooking time by 50-66%, thus saving both energy and time.

Low Hanging Fruit—appliances for food preparation

- ☐ Use the "lightest" energy-using appliance that will do the job.
- ☐ Avoid preheating the oven.
- ☐ When preparing a casserole or baking potatoes, preheat in the microwave for 3-5 minutes and reduce oven time by about a half.
- ☐ Use an electric kettle to heat water for beverages, unless making only one cup (heat in microwave).
- ☐ Avoid throwing food away. It has a negative impact on several levels: buying, cooking, refrigerating, and freezing food that will eventually be thrown out wastes energy. Prepare only enough for one meal or plan deliberate leftovers for a second.

Refrigerator-Freezer

Almost all models combine the refrigerator and freezer in one appliance and are powered by electricity. Some households also have a separate freezer for storing food bought in bulk or preserving the local summer harvest. The efficiency of a combination model is governed by age, size, configuration, and pattern of use (in that order). The average model has a life expectancy of 14-17 years and costs $800-$1,200 new. Take these and the following factors into account if you are considering replacing your refrigerator.

High Impact—Considerations for choosing a new refrigerator[25]

- ☐ It is time to replace if repairs will be more than $700 and/or it is more than 14 years old.
- ☐ A new refrigerator with an EnergyStar rating consumes 40% less energy than a conventional model from 2001 or later.
- ☐ A model with the freezer on the bottom uses 16% less energy than a side-by-side and 3% less than if the freezer is on top.
- ☐ Choose a size that will store enough staples that you use frequently and enough fresh food for your family for about week.
- ☐ If you are using an old refrigerator in the garage for extra storage, unplug it! Replacing your large, old model with a new small, unit

will pay for itself within a few years through the reduction in your electricity bills.

Low Hanging Fruit—using a refrigerator efficiently

- ☐ Set the temperature at 0º F for the freezer and 39-41º for the refrigerator portions.
- ☐ Follow instructions in the manual for regularly cleaning the coils on the back or bottom of the refrigerator.
- ☐ Remove all ingredients that you need for preparing a meal at once.
- ☐ Close the door/drawer promptly.
- ☐ Compost (preferably) or discard outdated food promptly to prevent spoilage and eliminate cooling food that is not wanted.
- ☐ Consume leftovers before they spoil.

Lighting

Although it represents only about 10% of the household energy budget, lighting is one of the easiest places to start tithing. Lightbulbs come in several varieties: traditional incandescent, halogen incandescent, compact fluorescent (CFL), and light emitting diodes (LED). Their energy use is measured in watts of electricity and the light output in lumens. LEDs are by far the most energy efficient,[26] their lifetime is the longest, and the quality of the light most resembles daylight. Halogen and fluorescent bulbs (whether compact or traditional shape) are toxic to the environment when not disposed of properly. They should be returned to a hardware store or town hazardous waste collection site.

Energy Used by Light Bulbs with Equivalent Lumens
https://smarterhouse.org

Incandescent	Light Output lumens	Halogen	CFL	LED
40	450	29	9-13	6-8
60	800	43	13-15	9-11
75	1,100	53	18-25	13-15
100	1,600	72	23-30	18-20
150	2,600		30-52	

Low Hanging Fruit—smart lighting

- ☐ Replace all screw-in lightbulbs with LED's. Dispose of CFLs responsibly.
- ☐ Convert ceiling fixtures from CFL to LEDs as the CFLs burn out (after 3-5 years).
- ☐ Turn off lights when not needed.
- ☐ Light the activity, not the room: more light is needed for reading than for conversation, and close-by lamps give more light than ceiling fixtures of the same wattage.
- ☐ Install enough lights to enable passing from one end of your house to the other, turning lights on and off as you go.
- ☐ Turn off outdoor lights when not needed or replace with ones that are only on in the dark.
- ☐ Replace outdoor lights on walkways with solar powered lanterns.

Electronic Devices

Most of us are so accustomed to the convenience of an array of electronic devices that we do not think about their carbon footprint and cannot conceive of doing without them. They consume an estimated 7-15% of our domestic electricity, but that does not include the energy consumed in their manufacture and distribution, nor that consumed by the large servers that maintain our instant access to a vast array of on-line information and purchasing. Considering that most of these devices either burn out or become outdated in five years or less, it is obvious that they represent a huge environmental concern that is growing faster than most of us can comprehend.

Most of the domestic energy consumption for our devices goes to our entertainment (via TV, DVR, music player, radio, etc.) and home office equipment (computers, Wi-Fi/modems, and printers), while other smaller devices, including a growing number of portables with battery chargers, account for the remainder. While none of these devices consumes a large amount of electricity by itself, the large and growing number that we own has a compounded effect on our consumption of electricity. This primer only addresses the basics of how we can responsibly manage and tithe our device use; for more information see, for example, the Smarter House website.[27]

High Impact Changes—electronic devices

- ☐ Observe an Electronic Sabbath day, i.e. a device-free day when you abstain from actively using TVs, computers, and cellphones.
- ☐ Only replace a device when it no longer functions and cannot be repaired.
- ☐ When buying or replacing a device consider the smallest, simplest model that will work for you, and a used one that has been reconditioned.

Low Hanging Fruit—smart use of devices

- ☐ Power down a device as soon as you are no longer using it or set to "sleep mode."
- ☐ Set your computer to turn off the display screen after 3-5 minutes of inactivity and the hard drive after 15 minutes.
- ☐ Consider whether your home entertainment components (TV, DVR, and music system) need to be in standby mode, i.e. whether it is necessary to be able to activate them with a remote control 24/7 or whether they should simply be completely turned off when not in use.
- ☐ Un-plug portable devices from the charger when the battery is recharged.
- ☐ Un-plug chargers from the electrical socket when not in us—they draw power even when no device is plugged in.
- ☐ Inspect your home for extraneous "on-lights," unnecessary clocks, dashboards, etc. and see if they can be disabled or turned off.
- ☐ Consider using "smart" power bars that monitor usage of devices and can turn off outlets, for instance, when there is a sudden drop in usage, such as when charging is complete.

Automobiles

The impact of the personal automobile on climate change is too complex a subject to explore fully in this primer and we will only touch on a few broad issues. Some of the topics covered may seem obvious but are worth mentioning to stimulate discussion and new ideas.

The amount of greenhouse gases that a car emits is a function of its size and power-source. In general, the power source ranking (highest to lowest impact) is: gasoline, diesel, hybrid, electric (EV or plug-in). This order is only approximate with the following caveats:

- Gasoline vs. Diesel: Though diesel powered cars have 25 to 35% better mileage than those powered by gasoline, they can emit 25 to 400 times more soot. Although soot suspended in the air has a shorter lifetime than carbon dioxide (CO_2), it has a greater immediate warming effect. A gasoline fueled car may therefore be a better choice in the short term.[28]
- Whether an all-electric car is preferable depends on the source of the electricity, i.e. how much of it is generated by burning fossil fuels, like coal and natural gas, as opposed to solar, wind, or hydroelectric (see section on electricity in Chapter *3).*

If you are contemplating purchasing a new or used car, it makes environmental sense to choose the car with the highest posted fuel-efficiency among those of otherwise comparable size and taking these caveats into consideration. Obviously, price is also a factor in a decision only you can make.

Keep creation in mind when you plan your vacations. Must you fly or travel to a distant place or would going closer to home be equally wonderful? A large airplane emits about as much CO_2e **per passenger** as driving the same distance in a standard car, but if you are taking the whole family and you drive a hybrid, the savings are obvious.

In addition to your choice of car, your driving habits are extremely important. Here are a few obvious suggestions:

Low Hanging Fruit—smart driving

- ☐ Reduce the amount you drive by: having a regular "no-car Sabbath day," car-pooling, combining errands, and reducing the frequency of grocery shopping.
- ☐ If you own more than one car, use the smallest and/or most efficient for the particular task at hand.
- ☐ Consider alternative modes of transportation according to the distance you need to go: walk, bicycle, public transportation, or train.

- ☐ If you have children, discuss how they can most responsibly get to school. As age-appropriate, teach them how to use alternative modes of transportation.
- ☐ Consider tele-commuting to work, when possible.
- ☐ Attend meetings by video conference or skype.
- ☐ Hone your driving technique.[29] The most important hints are:
  - ◊ Avoid jack-rabbit starts and stops.
  - ◊ Be easy on the pedal and observe the speed limit! If you travel 66 mph in a 60 zone you are reducing your fuel efficiency by about 10%. Plan ahead, give yourself enough time to get to your destination, and enjoy the scenery or radio! An easy way to tithe!
  - ◊ Use your cruise control on the highway if you are going more than a few miles in moderate or light traffic.
  - ◊ Schedule errands for off-rush hour or on the commute home.

Low Hanging Fruit—Vehicle Maintenance

- ☐ You have probably heard a hundred times to check your tire pressure, but how often do you do it? Well it is time to change that! Each PSI that your tires are underinflated will reduce fuel efficiency by about 0.2%.
- ☐ Make sure that your trunk and backseat are free from clutter. Each 100 pounds that is in your car will reduce fuel efficiency by 1-2%.
- ☐ If you are in the market for new tires, look at low roll resistance ones. They are more expensive but will increase fuel efficiency by several percent.

Best Suggestions

We re-emphasize the value of using the Primer in a group or workshop setting for mutual support, discussion, and exchange of information. For example, a master list can be brainstormed by the whole group or in break-out sessions.[30] During the first round, participants are free to make suggestions without comments or criticism from anyone, followed by a discussion period. During wrap-up, the list is winnowed or prioritized by some criteria of practicability or highest expected impact.

In the space below, jot down what you think are the 5-10 "best" or most practical suggestions from the lists or that arise during group discussion. Begin your tithing with these practices. Make a separate list of 1-2 higher cost appliances that you will explore for purchase in the near future. Make similar lists for your HOW to address.

End Notes for Chapter 2

---

[16] From the Breastplate of St. Patrick, http://www.worldprayers.org/archive/prayers/invocations/i_arise_today_through_a_mighty.html
[17] https://SmarterHouse.org
[18] www.interfaithpowerandlight.org or www.mipandl.org (in Massachusetts)
[19] https://www.masssave.com/en/saving/energy-assessments/
[20] EnergyStar.gov
[21] U.S. Green Building Council and LEED certification: www.USGBC.org
[22] https://water.usgs.gov/edu/qa-home-percapita.html
[23] https://www.epa.gov/watersense/how-we-use-water
[24] https://www.treehugger.com/clean-technology/ask-pablo-electric-kettle-stove-or-microwave-oven.html
[25] https://www.e-education.psu.edu/egee102/node/1993
[26] https://smarterhouse.org/content/choosing-right-bulb
[27] https://smarterhouse.org/appliances-energy/home-electronics
[28] https://news.stanford.edu/pr/02/jacobsonJGR1023.html
[29] https://greenercars.org/green-driving/driving-fueling
[30] https://en.wikipedia.org/wiki/Brainstorming

Chapter 3. TRACKING USAGE THROUGH UTILITY BILLS

O children of Adam, take your adornment at every masjid [place of worship], and eat and drink, but be not excessive. Indeed, He likes not those who commit excess.

Qur'an 7:31 [31]

Utility and gasoline bills are the "scale" on which we weigh the results of our carbon-diet and "weigh-in" to hold ourselves accountable for our tithing efforts by tracking our usage. Just as weight varies from day to day, the use of a utility will vary with circumstances like daily fluctuations in the temperature or changes in season. Although we can compare our usage month-to-month the whole picture will not be clear except over a long period of time. Therefore, we set our goals for the current year at 10% less than the goal for the previous year, however outside factors in addition to our changing habits may affect whether we achieve or exceed our goal by the end of the year. To set your goal for the first year, make a best guess if you do not have data for the previous year (sometimes the data can be found on the bill or on the utility's website).

We will examine each utility in turn. At the end of the chapter there are templates of tables for simple tracking, which you can freely copy or visit www.creation.tithe.com to download electronic versions. The first template can be used to track utilities for which you receive monthly bills (like electricity and natural gas), quarterly (like water bills), and those that are episodic, (i.e. you pay for propane or heating oil as it is delivered). The second template is useful for tracking gasoline consumption. It is slightly different in that you have to record the miles from the odometer each time you fill your tank. You or your group can also use these templates for designing your own spreadsheets to make the calculations automatic. Appendix I contains two partially completed spreadsheets used to track natural gas and gasoline during the early months of 2018 compared to 2017. Spreadsheets that automate much of the process of tracking utilities can be downloaded from the internet, for example from the Massachusetts Interfaith Power and Light website.[32] These spreadsheets have more extensive analyses, including taking account of the weather by using data during the month on the average temperature, or the number of heating/cooling days. See the website for definitions.

Tracking Spreadsheets

Mass Interfaith Power & Light has comprehensive tracking Excel spreadsheets on their website: www.MIPandL.org.

To download (free), go to the Resource page and click on "Utility Use & Cost Spreadsheet for":

- *Individuals*
- *Congregations*

Electricity ϟϟϟ

Electricity that is generated without the use of fossil fuels is the energy source of the future. It is not too soon to start preparing for that future by understanding how utility companies currently generate and distribute power.

Your electric bill probably consists of two charges: a *generation charge* and a *delivery charge,* each in proportion to the amount your household uses, measured in *kwh* (kilowatt-hours). Your bill will come from the company that delivers, but the generation charges may come from a different company and even on a separate bill.

Electricity is generated from a mixture of sources: nuclear power, fossil-fuels, (i.e. coal, oil, and natural gas), wood, garbage, wind, solar, or hydro-electric. The last three are often called *renewables,* but *sustainable* is a better term. In Massachusetts, the primary source is natural gas.[33] You may be able to choose a different supplier that generates a proportion from sustainable sources, though usually at a slightly higher price. Be aware that some companies that claim to generate all or a portion of their electricity from "renewable" or *fossil-fuel-free* sources, may be burning wood scraps or garbage (yes, they are certainly renewable but hardly free of greenhouse gas emissions). In Massachusetts, you can support the generation of electricity from completely sustainable resources, for example by joining the Energy Alliance.[34] Your usual utility company still delivers and issues the bills. Tracking is simplified if you sign up for paperless billing, e.g. Eversource.com in the Boston area.[35]

We will not dwell on the safety concerns with nuclear power plants nor with the huge environmental disaster represented by the use of coal. In Massachusetts, coal-fired power plants are gradually being phased out (we hope). Even hydro-electric power has a bad reputation among environmentalists because of its (past) dependence on huge dams that disrupt the eco-system of rivers and their watersheds. Newer dams have a much lower environmental impact[36] and there are efforts to generate electricity from ocean tides.

You probably will not have a choice of the company that delivers your electricity. It is likely part of the National Grid, which stabilizes the supply and you pay to maintain. The U.S. Energy Information Administration (EIA) estimates that about 5% of electricity is lost in transmission,[37] more in some areas where it needs to be transmitted over long distances.

If you or your HOW have installed *solar panels* (or are planning to), the power you generate is probably fed into the grid and you get your electricity from there,

Bishop Cederholm blesses a church solar array.

including when the sun isn't out. What you generate offsets your usage and you may pay nothing for your electricity by, what is called, "net-metering". It is important to understand that even if you generate your own electricity and your cost is 0, **tithing is still important** as it will reduce the CO_2 emitted into the atmosphere by the generating plant and reduce society's dependence on fossil-fuels.

If, despite all your best efforts, you do not seem to be cutting down on your kwh, it may be that you have switched some of your use from other utilities to electricity. For example, you may have reduced your gas consumption by using your big gas-powered stove less in favor of a more efficient small electrical toaster oven, microwave, or slow-cooker. Similarly, if you bought a plug-in automobile, you will be saving gasoline, but increasing your consumption of electricity.

Gas and Propane

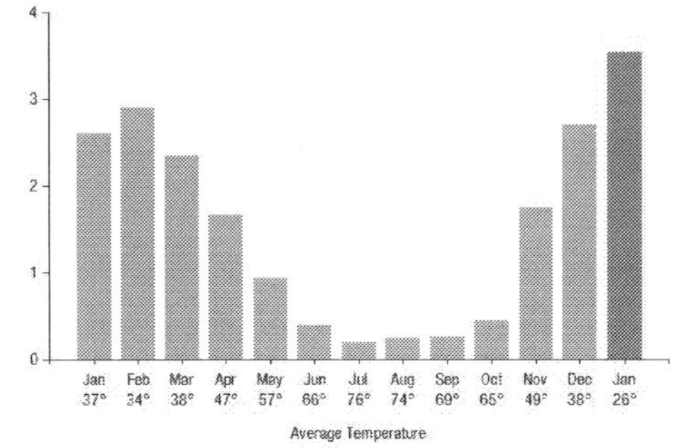

Natural Gas Usage Changes with Season and Temperature

Natural gas is commonly used in power plants for the generation of electricity. It burns "clean" in that mostly CO_2 is released, as opposed to burning coal which also releases soot (particle pollution), methane, and other greenhouse gases. For this reason, gas is sometimes considered a "transition" fuel, that will protect the environment until sustainable forms of energy can be developed on a scale that will meet society's needs. At present, many households and HOWs depend on gas for heat, hot water, and cooking as a less costly alternative to electricity. Billing is done according to the number of *therms* used. You can sign up for paperless billing on your utility company's website (e.g. Eversource.com in the Boston area), thus simplifying tracking as well as saving trees used to manufacture paper.

The "natural" part of gas is that mixtures of mostly methane and other carbon gases occur in pockets underground. The wholly unnatural part lies in the extraction process of "fracking," which involves forcing water and chemicals at high pressure into those pockets to release the gas, often causing earthquakes, contaminating surrounding aquafers, and draining ground water. A second problem arises in transmitting the gas through ageing and leaky pipelines thereby significantly reducing the amount of gas that is delivered to consumers and possibly killing urban trees with the poisonous fumes.

Propane, is an alternative to natural gas for houses, mobile homes, and campers that do not have access to piped municipal natural gas and also for cooking on outdoor grills. Propane is chemically related to methane and is a byproduct of refining oil from petroleum or natural gas processing, although some pockets do occur in nature.[38] It is delivered to homes in large cylinders. Billing is periodic and based on gallons delivered. Smaller portable tanks of propane for campers and outdoor grills are picked up in by the consumer, but we will postpone the discussion of these other uses until the next chapter.

Units of Heating Capacity for Gas and Oil

Heating capacity: *BTU or* British Thermal Units

1 BTU = heat needed to raise 1 lbs. water 1 F°
Natural gas: 1 *therm* = 100,000 BTU
Propane: 1 *gallon* = 0.9 therms or 91,502 BTU
Heating oil: 1 *gallon* = 138,500 BTU

Heating Oil

The oil used for heating is similar to diesel fuel for cars[39] and has the same profile of concerns about its hazard and impact on the environment. It is delivered direct to the house or HOW and stored (usually in the basement) in large tanks. The billing is periodic and based on gallons delivered. If practical, consider replacing your oil-furnace with a modern efficient gas-fired boiler or a complete HVAC system. Low or zero interest loans may be available and the considerable savings in your utility bills will help offset the payments.

Water

The average person in Massachusetts consumes 90 gallons of water each day, a shocking amount, especially when you multiply that figure by the number of people in a household. The picture is much worse for those who live in dry western parts of the U.S. where the average consumption is up to 200 gallons and that water must, in some cases, be pumped for hundreds of miles over the mountains.

Puerto Rico and Hurricane Maria

Hurricane Maria: Landfall 9/20/2017, category 4, 155 mph winds

Immediate Consequences: power grid effectively destroyed

Lack of services:	Electricity	Tap Water	Cell phones
After one week:	95%	<50%	95%
After one month:	88%	29%	40%
After three months:	45%	14%	10%

In most of the U.S. and Canada, we take for granted the availability of water that is safe to drink, but the supply is particularly sensitive to damage to the electrical grid caused by severe weather events. Climate change has continuing impact with increasing frequency and severity of crippling droughts, devastating floods, and monstrous hurricanes. During 2017, the droughts and wildfires in the western states and series of hurricanes along the Atlantic states and Caribbean Islands demonstrated the fragility of our water systems.[40]

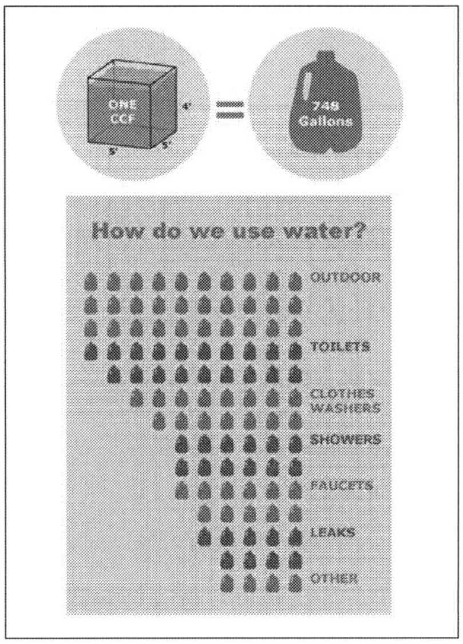

Most municipalities send bills periodically and give usage in CCF (aka HCF meaning Hundred Cubic Feet). To make this more meaningful, you can translate the CCF on a water bill into gallons by multiplying by 748, as in the graphic on the left.[41]

Although delivering water, processing so it is safe to drink, and disposing of the resulting sewage is not a large part of our carbon footprint, we are draining this increasingly scarce resource from ground water, rivers, and reservoirs.[42] Tithing is definitely worthwhile in raising your awareness and that of your community of how we use and abuse this essential gift from our Creator. In addition, examining your energy and water tracking records can help diagnose which of your efforts are the most effective and may inspire you to advocate for stronger municipal, state, and federal laws.

Gasoline and Diesel

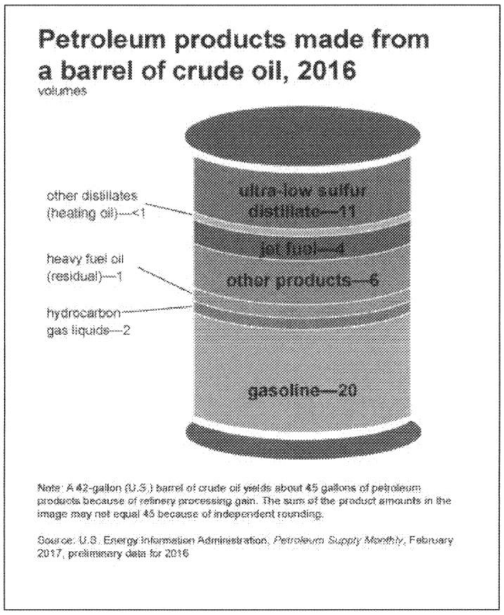

Gasoline comes from the refining of crude oil and represents about 45% of the petroleum products produced.[43] According to the Union of Concerned Scientists, cars and trucks account for about 20% of all emissions of greenhouse gases in the U.S.[44] Of course, this percentage includes the trucks (mostly diesel powered) that haul our food and goods across the country. For a discussion about how to drive more conscientiously and a comparison of gasoline and diesel-powered automobiles, see Chapter 2. If you travel a great deal by air, consider tracking your air miles separately. The amount of fuel burned per passenger by a large jet is about equivalent to driving an automobile the same distance. Chapter 4 contains a discussion of the environmental impact of forms of transportation other than personal automobiles.

Setting your Goals

Templates for tracking are at the end of this section, but before you start, you must set your tithing *goal* for each utility for the coming year.

Setting your goal for the first year. If you have records of usage for the previous year or can get them from your utility company's website, you simply reduce the total Units used the previous year by 10% (multiply by 0.90). Since the total billing period may differ by a day or two from year to year, it may be more accurate to reduce your total units per day (Units/D). If you have records for more than one year, you may want to compute the average to even out the effects of changeable weather (extra hot summers or very cold winters). If you do not have complete records, you may be able to infer your usage from the cost in your checkbook register, bank statement or credit card and the charge per unit of your utility company.

Setting your goals for subsequent years can be done by either: (A) reducing your actual **total usage** or usage/day from the previous year by 10%; or (B) reducing your **goal** from last year by 10%. Option B allows you to "carry" over the effect of a major switch to lower emission HVAC system or a more efficient automobile to subsequent years.

Template 2, used for tracking how much gasoline you consume, has a slightly different format since you have to record your car's odometer reading when you are at the pump. The table is open ended and the number of rows you need will depend on how frequently you refuel your car. You will need to track gasoline for each car that you drive. As before, you will set your goal for this year from your records for the previous year. For the first year you may have to make an educated guess, e.g. from the odometer reading of your car at the beginning and end of last year and its average mileage. If you drive a camper for vacations, you may want to complete a similar table for this vehicle in addition to your car.

Although Creation Tithing is not about saving money, the tables also allow you to track the cost. You may want to set aside the money saved from picking the low hanging fruit to help fund the cost of replacing energy gulping equipment, such as an outdated heating system. This may be an important consideration for your HOW in order to justify expenditures to the Board.

Many people find it difficult to interpret the bills and some of the computations daunting; therefore, we recommend sharing tracking skills within your group. If you have teenagers in your household, encourage them to participate. Including young people is an excellent way to involve them in preparing for what is, after all, their future. 😀

Tracking Templates ✎✎✎

You will find templates 1 and 2 for tracking your utility usage on the following two pages. You may copy these freely or download them from www.creationtithe.com. Template 1 is for tracking utilities from the bills you receive from the supplier showing your usage since the last billing period. The columns are labelled alphabetically and explanations for each given below the table, including which numbers you transpose directly from the bill and which you may have to calculate. The table has 12 numbered rows, corresponding to the number of months in a year. For utilities, such as water, propane, and heating oil, for which you do not receive a bill every month, just complete the row for the relevant month. You will need a copy of the template for each utility that you use and for each year that you tithe.

Template 1 for Tracking Utility Bills

An example of populated tables for Natural Gas in 2017 and 2018 are in Appendix I

A	B	C	D	E	F	G	H
Year(s)		**From Utility Bills**				**Calculate**	
MNTH	**DATE**	**DAYS (D)**	**UNITS**	**%DIFF**	**$**	**Units/D**	**Av % Diff**
1							
2							
3							
4							
5							
6							
7							
8							
9							
10							
11							
12							
TOTAL							
Goal							

Overall: Use the template for each of the utilities you use for every year of your tithing, including the year before you start (if available). If your start month is other than January, the table will overlap two years. Enter the numbers from you bills into Columns A-F, and calculate Columns G-H.

Column A: Month 1 is your start month and may not be January. For utilities that are not billed monthly, simply fill in the row of the month the meter was last read, or the tank refilled.

Column B: The date that the meter was read, or the tank refilled (not the date of issue or due date).

Column C: The number of days since the last bill. If not listed on the bill you have to count.

Column D: The units of usage measurement, *KWh* for electricity, *Therms* for natural gas, *gallons* for heating oil and propane. If the usage on the water bill is in CCF or HCF, convert to gallons by multiplying by 748.

Column E: The % difference in usage from the previous year for this month may be on the bill. It is affected by the difference in weather as well as changes in your habits.

Column F: Decreasing your cost is not the goal of tithing but saving money might be important.

Column G: Divide Col. D by Col. C. This too will vary considerably with the weather.

Column H: For an easy estimate simply average the % difference in Col. E over the number of months to date. OR, to be precise, compute the *weighted average* using days as the weights.

TOTAL row: Enter the sums of columns C, D, and F and compute G & H as before.

GOAL row: Set your tithing goal for the coming year before you begin tracking, instructions for the first year and for subsequent years are given in the previous section on **Setting Your Goals.**

✎✎✎Tracking Template 2, for Gasoline Usage ✎✎✎

An example of a partially populated table for gasoline is in Appendix I

A	B	C	D	E	F	G	H
At the Pump				**Calculate**			
Date	**Odom**	**Gallons**	**Cost**	**Days**	**Miles**	**Gal/Day**	**M/Gal**
		Fill Tank	xxxxxx	xxxxx	xxxxx	xxxxx	xxxxx
Etc.							
TOTAL							
GOAL							

Overall: Use one table each year of your tithing and one for each car in your household. Depending on how much you drive and how often you fill up, you may need a longer table.

Columns A-D: Hint: make note the entries for Columns A-D while you are at the pump and filling your tank. To get started, record just the odometer reading and date for a baseline. As with utilities, tracking cost (Column D) is optional and not the focus of your tithing.

Columns E-F: Calculate these by comparing the odometer readings and dates to the previous row, respectively. If the number of miles since the last fill-up is about twice as large as usual, you may have accidentally skipped a reading. In which case, you should guess, e.g. by doubling the number of gallons and the cost (Col. C and D) and leaving the missing odometer reading blank. ☺

Column G-H: Calculate Col. G by dividing Col. C by Col. E to see your average consumption. Your *mileage* is in Col. H (divide Col. F by Col. C). Both will vary with your highway vs. city driving.

TOTAL row: The numbers in Col. C, D, E, and F in this row are simply the totals of each column. Compute Col. G and H as before. You may want to do this after each refill, to monitor your progress.

GOAL row: Set you tithing goal for the coming year before you begin tracking. Instructions for the first year and for subsequent years are given in the previous section on **Setting Your Goals.**

End Notes for Chapter 3

[31] http://www.hebron.edu/quran/mobile/translate.php?l=2
[32] http://www.mipandl.org/faith_res.html
[33] https://www.eia.gov/electricity/state/massachusetts/
[34] https://www.massenergy.org/renewable-energy/nstar
[35] https://www.eversource.com/content/ema-c
[36] https://lowimpacthydro.org
[37] https://www.eia.gov/tools/faqs/faq.php?id=105&t=3
[38] https://en.wikipedia.org/wiki/Propane
[39] https://en.wikipedia.org/wiki/Heating_oil
[40] https://en.wikipedia.org/wiki/Hurricane_Maria
[41] https://www.epa.gov/watersense/understanding-your-water-bill
[42] http://leansixsigmaenvironment.org/index.php/how-much-impact-does-one-gallon-of-water-in-your-home-or-business-have-on-the-environment/
[43] https://www.eia.gov/energyexplained/index.cfm?page=oil_refining
[44] https://www.ucsusa.org/clean-vehicles/car-emissions-and-global-warming#.WnnFNCOZPo1

Chapter 4. OTHER CONSIDERATIONS

Our beloved Earth, so touchingly looked upon as the Universal Mother, has nurtured mankind through millions of years of growth and evolution. Now centuries of rapacious exploitation of the planet have caught up with us, and a radical change in our relationship with nature is no longer an option. It is a matter of survival. We cannot continue to destroy nature without also destroying ourselves. Hindu Declaration on Climate Change, 2009 [45]

Creation Tithing is not easy. However, whether or not we, as a global society, take our moral responsibility seriously, the fact is that our lifestyles and consumer habits will drastically change in the near future as the impact of climate change becomes a reality and not just a threat from our environmental prophets. We can either grasp the nettle of change and meet the challenge with positive action or wait until we have no choice. The resources that we now blithely take for granted will no longer be there and we will have no alternatives waiting in the wings. As the climate worsens, we will face more and more weather disasters and our lives will be threatened by growing civil unrest and persistent wars as economies, farms, and waterways are laid waste.

No one claims that a few concerned individuals can achieve a reversal of climate change with the simple act of cutting back on their consumption. But hear what the Holy Spirit is saying to the people, listen to what you are called to do, and let the Creator lift your efforts to the mountain top.

This chapter will discuss Creation Care issues beyond simple tithing: what to do if you fall short in your tithing effort, purchasing carbon offsets, other consumer goods that have major impacts on greenhouse gases in the atmosphere, and advocating for change at your place of work or in the public sphere.

Falling short of achieving your tithing goals

There are many reasons why, despite your best resolution and sincere efforts, you may not be able to achieve the goal of reducing each of your billable utilities by 10% the first year, never mind in each of the subsequent years. In the table, below we list just a few obvious possible barriers to tithing. These are very general, so it might be helpful if you add specific details of your particular circumstances that might explain a short-fall. You may also wish to add two or three of your own.

Common Barriers to Tithing

- ☐ I find it difficult to convince myself or the other members of my household that we can really make a difference.

- ☐ I/we have already picked the low-hanging fruit and made all the higher impact changes in our appliances that we can make or can afford.

- ☐ I do not have the time (or money) for tithing to be a high priority.

- ☐ I /we have made a real effort, but one or more of my utilities still fell short of my goal. Explain which utility(ies) and why that may be so:

It is not our intention to make an exhaustive list for each of us has different circumstances and constraints on what we are able to do, but we ask you please to consider and discuss each of the items with your household and in your Creation Tithing group. Do not use your barriers as an easy excuse, but honestly and prayerfully consider what you might do to increase your tithing or work in other ways to mitigate climate change.

What to do? Again, there are several obvious simple remedies:

Low Hanging Fruit—Making up Shortfalls

- ☐ *Do the best you can.* The 10% goal is not a magic number; it is the effort, sharing, and witness that count.

- ☐ *Set a long-range goal.* A goal of 10% each year, is an overall reduction of about 40% after 5 years. Set you goals using option B in Chapter 3 and work towards achieving that.
- ☐ *Compute your reduction in total CO_2* over all your utilities by translating the total units of consumption for each of your utilities into "greenhouse gas equivalents."[46] You may exceed 10% in some areas and not in others, but your total greenhouse gas consumption may still be on target.
- ☐ *Review Chapter 2.* Perhaps there is something you have missed, or way will open for you to try something new.
- ☐ *Help others achieve their goal.* If you have already cut back as much as you can, perhaps your mission is to help others and provide expertise, support, and experience to your tithing group. Start a Creation Care blog, Facebook page, or email circulation list to publicize what your group is doing.
- ☐ *At your HOW.* Advocate for Creation Care in your worship and for lessening the carbon footprint of its property and activities. Volunteer to track tithing for your HOW.
- ☐ *Advocate for change in other places.* Each individual can only make small changes and it is important to begin with yourself, but in the long run it is essential that society as a whole embrace the need for reform of consumption at places of business, in public buildings, and in public policy. We discuss a few ideas in the last two sections of this chapter.
- ☐ *Purchase carbon offsets*—see the next section.
- ☐ *Reduce your carbon footprint in other ways.* Examine your consumption of other products, even if you can't track them in any obvious way—see the section on Other Consumer Products for a few ideas.

Carbon Offsets and Credits

The goal that was set at the beginning of this primer was to reduce our greenhouse gas emissions by 10% per year. With the suggestions given in this primer, meeting this goal for the first few years should be achievable, but what happens after that?

What if you have done everything that you can think of but can no longer meet the tithing goal?

Beyond utilities and gasoline usage, there are countless other ways to reduce your carbon footprint and there are several great websites and books that have suggestions on actions. Appendix II has a listing of some of these resources.

One way to make up for any shortfall that you may have is to contribute to greenhouse gas reduction projects. These projects can be located anywhere on the planet and can include carbon sequestration, methane reduction, and increasing energy generation efficiency. You may think that participating in these is difficult but through several "Carbon Credit" programs it is very easy to do. Through these programs you purchase a carbon offset, which means you are buying a fractional amount of a project, somewhere in the world, that will reduce the amount of carbon put into the atmosphere.

Carbon offsets have three common characteristics. The year that the carbon reduction will take place, the type of reduction, and the certification that the offset is really going to do what it said it was going to do. The certification of the carbon offset with an independent audit is important since the market is not regulated. Examples of certification include: The Clean Development Mechanism (CDM), The Gold Standard and Gold Standard Verified Emission Reduction (VER).

Offset Certification

VER+
Robust. Reliable. Verified.
Gold Standard
Climate Security & Sustainable Development

Several internet sites allow you to purchase certified carbon offsets all at once or with monthly payments:

- Through Terrapass (*www.terrapass*.com) you can purchase Carbon Offsets for $5.95 per 1,000 lbs, Renewable Energy Credits for $5 per 1,000 MWatts, or BEF Water Restoration Credits® for $2 per 1,000 gal.[47]
- Similarly, in Canada, offsets can be purchased from Less Emissions Inc., including for emissions for air-travel.[48]

Though the numbers vary greatly depending on location, lifestyle, and type of dwelling, the average American has a carbon footprint of about 22 tons. Of this, about 60%, or 13.2 tons, is related to heating, electricity, and transportation. If you wanted to do a 10% reduction in CO_2 and you have three people in your household, it would cost you about $400 to purchase carbon offsets for that amount.[49]

Other Consumer Products.

In addition to usage of utilities and gasoline that we can track through our bills, most of the stuff we buy has a major impact on atmospheric concentrations of CO2e (carbon dioxide equivalent). We will only mention a few, but please refer to more complete discussions in the suggested references of Appendix II. The major guideline is to learn to appreciate the gift to live simply.[50]

- *Food.* The carbon footprint of producing and transporting the food we eat is a HUGE topic and well beyond the scope of this primer. A few rules of thumb:
  - ☐ Buy seasonal produce from local farms whenever possible.
  - ☐ Plant a victory-garden or join a local CSA (Community-supported-agriculture).
  - ☐ *Preserve food.* Can, dry, or freeze local produce when it is season.
  - ☐ *Choose organically grown produce.*
  - ☐ *Reduce* (or eliminate) your consumption of meat, especially of beef and pork. Look for grass-fed beef.
  - ☐ *Avoid food with many travel-miles.* Beware of buying foods that have travelled thousands of miles before reaching you.
  - ☐ *Study food production* methods and avoid those foods that have heavy carbon footprints.
  - ☐ *Avoid heavily processed foods.*
  - ☐ *Avoid wasting food.*
- *Clothing.* The source of most clothing is a mystery of complex supply chains and foreign manufacturing, sometimes produced under unacceptable working conditions. The best advice is: buy less and simplify your wardrobe.
- *Reduce, reuse, recycle.* The most important of these is "reduce," but the other two should also be an integral part of every climate-conscious citizen's life. In addition, consider composting, refilling, and remanufacturing possibilities.
  - ☐ *Donation Store.* Bring unneeded household items and clean wearable clothing to a donation or "thrift" store.
  - ☐ *Recycle responsibly.* In addition to paper, plastic, and aluminum cans, bring spent LED, fluorescent tubes, and leftover paint to your municipal hazardous waste facility. Some suppliers will also help you recycle.
  - ☐ *Compost your food waste. If you have a garden build a composting bin for food and garden waste and produce your own mulch. Some municipalities have curb-side composting pick-up.*
  - ☐ *Refilling.* Toner cartridges can be refilled rather than discarded.

- ☐ *Remanufacture.* Used electronic equipment can often be remanufactured and reconditioned at much lower energy cost and less use of scarce materials than if produced from scratch.

Green Buildings.

Whether building or planning a major renovation of your house or HOW consider "building green" using LEED certified contractors and building materials. More information can be found on the web-site of the U.S. Green Building Council, especially on LEED (*Leadership in Environmental Design)* certification programs.[51] The increase in cost for certification may be smaller than you expect and will save you money in the long run, especially if you become eligible for low interest, energy-efficiency loans. Remember that "small is beautiful" as well as more energy efficient.[52]

Advocating for change at work.

Some work places have active sustainability programs and have enacted changes that have significantly reduced their carbon footprints, but others are surprisingly resistant to change especially since many of these changes can be very low cost as well as carbon effective. Join or form a Green Team to apply what you have learned to your place of business in utility savings. Look into purchasing practices that could lighten the business footprint, encourage changes in the way employees commute to work by facilitating use of carpools, bicycling, and public transportation. Examine the feasibility of changing work hours to off rush-hour commuting, four-day work weeks, and/or telecommuting. Reduce company travel by encouraging tele-conferences. The possibilities are endless and may have a large impact. For example, Lean and Six Sigma methodologies helped reduce the carbon footprint of Portland Oregon businesses. You may find their website useful, including a bunch of free how-to materials to download.[53]

Advocating for change in public policy.

Creation tithing is the means to the revolutionary end of the misuse and abuse of our island home. But while individual and group efforts are important, without dedicated effort to change policy in the public sector, that revolution will not happen. It is the moral duty of God's people to join with their neighbors in acting by demonstrating, marching, and signing petitions. Write letters to legislators, city officials, and your local newspaper. Donate money and time to one or more "frontline" climate activist groups; some of the most prominent are: 350.org, Union of Concerned Scientists, Mothers Out Front, GreenFaith, and The Environmental Defense Fund; these and others are listed in Appendix II and chapter end notes.

Many climate activists agree that our best hope for curbing fossil fuel emissions is the enactment of laws that put a price on carbon emissions on the international, federal, or even state levels. Some hold out hope that technological innovation will mitigate the climate crisis by diverting our energy gulping habits to sustainable life-styles. But we are not there yet, so while we advocate for that change, do not be a hypocrite, "first take the log out of your own eye, and then you will see clearly to take the speck out of your neighbor's eye."[54] Do not be shy about what you are doing, but shout it from the mountain tops. Let your Creation Tithing witness your determination and become an "outward and visible sign of an inward and spiritual grace."[55] Let us not just advocate for mitigating climate change but work to change the social and political climate.

End Notes for Chapter 4

[45] Hindu statement on Climate Change, presented at the Parliament of the World's Religions, Melbourne, Australia, 2009 As quoted by Matthew McDermott in "Our Sacred Earth, Hinduism and The Earth," *Hinduism Today Magazine,* Web Edition, April/May, June 2011 https://parliamentofreligions.org/content/parliament-worlds-religions-2009-0.

[46] https://www.epa.gov/energy/greenhouse-gases-equivalencies-calculator-calculations-and-references

[47] www.terrapass.com

[48] www.less.ca

[49] http://shrinkthatfootprint.com/american-carbon-footprint

[50] ***Simple Gifts*** by Joseph Brackett for the Shaker Community at Gorham, Maine. (1848)

[51] https://new.usgbc.org and https://new.usgbc.org/leed

[52] EF Shumacher, *small is beautiful, a study of economics as if people mattered,* Blond & Briggs (1973)

[53] http://leansixsigmaenvironment.org/index.php/how-much-impact-does-one-gallon-of-water-in-your-home-or-business-have-on-the-environment/

[54] Bible (NSRV), Matthew 7.1-5

[55] Book of Common Prayer (Episcopal) p. 857 and https://www.episcopalchurch.org/page/sacraments

Appendix I. Tracking Examples

Tracking Natural Gas—Example

A	B	C	D	E	F	G	H
2017		**From Natural Gas Bills**				**Calculate**	
MONTH	**DATE**	**DAYS (D)**	**THERMS** (Th)	**%DIFF**	**$**	**Th/D**	**Av % Diff**
JANUARY	1/23/17	33	84	NA	NA	2.5	NA
FEBRUARY	2/21/17	29	82	-	-	2.8	
MARCH	3/22/17	29	66	-	-	2.3	
APRIL	4/21/17	30	49	-	-	1.6	
MAY	5/22/17	31	28	-	-	0.9	
JUNE	6/21/17	30	12	-	-	0.4	
JULY	7/21/17	30	6	-	-	0.2	
AUGUST	8/22/17	32	8	-	-	0.3	
SEPTEMBER	9/22/17	31	8	-	-	0.3	
OCTOBER	10/24/17	32	14	-	-	0.4	
NOVEMBER	11/21/17	28	48	-	-	1.7	
DECEMBER	12/21/17	30	79	-	-	2.6	
TOTAL/AV.		365	484			1.33	
Goal			NA			NA	

A	B	C	D	E	F	G	H
2018		**From Natural Gas Bills**				**Calculate**	
MONTH	**DATE**	**DAYS (D)**	**THERMS (Th)**	**%DIFF**	**$**	**Th/D**	**~Av % Diff**
JANUARY	1/23/18	33	113	25.7		3.4	25.7
FEBRUARY	2/21/18	29	65	-26.2		2.2	-.3
MARCH	3/21/18	29	58	-13.8		2.0	-4.8
APRIL	4/23/18	32	53	1.38		1.66	-12.9
MAY							
JUNE							
JULY							
AUGUST							
SEPTEMBER							
OCTOBER							
NOVEMBER							
DECEMBER							
TOTAL/AV.							
Goal (90% of 2017)			435.6	-10.0		1.20	-10.0

N.B. January was an unusually cold month

✎✎✎Tracking Gasoline Usage—Example ✎✎✎

A	B	C	D	E	F	G	H
At the Pump				**Calculate**			
Date	**Odom (miles)**	**Gallons**	**Cost**	**Days**	**Miles**	**Gal/Day**	**M/Gal**
10/28/17	141	Full Tank	xxxxxx	xxxxx	xxxxx	xxxxx	xxxxx
1/3/18	419	9.9		67	278	0.15	28.1
2/21/18	753	10.1		49	334	0.21	33.1
4/19/18	1035	8.5		57	282	0.15	33.2
5/27/18	1503	9.9		38	468	0.26	47.3
Etc.							
TOTAL/Av. (to latest fill)		38.4		211	1,362	0.18	35.5
GOAL (guess)						0.20	30.8

N.B. Car (Prius, Hybrid) was new in October 2017, the goal for 2018 was estimated from incomplete data. In 2018, most driving was stop-and-go in the city.

Appendix II. SUGGESTIONS FOR FURTHER READING

There is a wealth of information available on the topics of climate change and sustainable living. However, just like with journalism, there are also many resources that have "fake" news and information. When looking at a new source, cross check the information with a known, reliable source. The resources below have been extensively used and can be considered trusted sources.

Resources on Climate Change

- ***Climate Church, Climate World: How People of Faith Must Work for Change,*** Jim Antal (2018)
- ***Encyclical on Climate Change and Inequality: On Care for Our Common Home,*** Pope Francis (2015)
- ***No is Not Enough: Resisting Trump's Shock Politics and Winning the World we need,*** Naomi Klein (2017)
- *This Changes Everything: Capitalism vs. the Climate,* Naomi Klein (2014)

Resources on Sustainable Living

- ***Cooler Smarter: practical steps for low-carbon living: expert advice from the Union of Concerned Scientists,*** Seth Shulman, et al., Island Press (2012)
- ***Feasible Planet, a Guide to More Sustainable Living***, Ken Kroes, *(2017)*
- ***Low Carbon Diet: a 30-day program to lose 5,000 pounds,*** David Gershon, Empowerment Institute (2006)

Sacred Scripture

- *Bible:*
    - ***New Revised Standard Version*** *(NRSV),* Anglicized Edition, *National Council of the Churches of Christ in the United States of America. (1995)* www.nrsvbibles.org
    - ***Tanakh*** *(Torah, Nevi'im, and Ketuvim), English translations by The Jewish Publication Society, (1907, 1963, 1985)*
    - ***The Voice***™*, Thomas Nelson, translator, Ecclesia Bible Society (2008)*
- *Qur'an:*
    - ***Quran in English,*** Talal Itani, translator (2017). **www.clearquran.com**
    - ***The Noble Qur'an in the English Language****,* Dr. Muhammad Taqi-ud-Din Al-Hilali and Dr. Muhammad Muhsin Khan, translators. http://www.hebron.edu/quran/mobile/translate.php?l=2

Useful Websites for Sustainable Living

- **www.EnergyStar.gov**
- **www.grist.org**
- **www.interfaithpowerandlight.org** (aka. IP&L)
- **www.macucc.org/carbonfast**
- **www.mipandl.org**, the Massachusetts node of IP&L
- **www.SmarterHouse.org**, *a project of the American council for an Energy-Efficient Economy*
- **www.treehugger.com**
- **www.USGBC.org -** U.S. Green Building Council and LEED certification

Acknowledgments

Creation Tithing—a Primer owes its completion to a serendipitous online meeting between authors Ken Kroes and Anne Ipsen when we discovered a shared passion for mitigating climate change and a belief that the best hope for a civilized future is for people of faith to act together. If climate change is the moral challenge of our time, then religious communities are perfect agents for change towards climate justice.

We wanted the Primer to be practical along with having a firm spiritual foundation and are therefore grateful that Rev. Lise Hildebrandt infused the writing, especially of the Introduction and Chapter 1, with her special insight. We also wanted to embrace other faith traditions and appreciate Rabbi Katy Allen's willingness to not only explain the roots of tithing in the Torah but to endorse the Primer. The Boston Islamic Seminary helped us better understand Islamic *Zakat* and amazed us with the spiritual practice of *Sadaqa*, the *giving up* of a blessing from Allah; this is exactly the message of the Primer.

Fruitful as our collaboration has been, we could not have succeeded without the contribution of many other people, only a few of whom can be named here. For many years, The Rt. Rev. Bud Cederholm, the "Green Bishop" of the Episcopal Diocese of Massachusetts, has been an inspiration with his devotion to Creation Care and call to worshipful action. He and the Diocesan Creation Care Network embraced the concept of the Creation Tithing Program from its inception and have enthusiastically supported the Primer. Individual members of this team of volunteers have been especially helpful. Jennifer Garrett corrected a myriad of typos, format errors, and grammatical blunders with her thorough copy editing. Dawn Tesorero, is helping to publicize the Creation Tithing Program and the Primer through newsletters, emails, and social media. Jim Nail and Thomas Nutt-Powell contributed environmental expertise and taught us language that communicates across false religious barriers. They assured us of continuing collaboration with Massachusetts Interfaith Power and Light as they light the path of "Faith in Action for Creation."

Rev. Dr. Margaret Bullitt-Jonas, Missioner for Creation Care has for many years led the environmental movement with her faithful presence at conferences and demonstrations, and at the helm of spiritual workshops and retreats. We therefore owe this busy woman special thanks for her endorsement of the Primer and advice about reaching out to other faith traditions.

We dedicate the Primer to our spouses, Linda and Jay, who have graciously "tithed" their expectation of our companionship so that we could attend to our passion for the care of creation.

About the Authors

Anne Ipsen is a writer, speaker, and environmentalist. She is the author of two memoirs and five novels. All have historical themes, except the latest, *Green Valley, a Climate Change Novel,* which takes place in the near future and describes a rural community that models a resilient way of life as New England leads the Carbon Revolution.

Before leaving academia in order to write full-time, Dr. Ipsen was Professor of Public Health at the University of Minnesota. More recently, Anne's interests have turned to the environment and the looming disasters associated with climate change. She started working on divestment from fossil fuel investments, first through her church and then with the Creation Care Network of the Episcopal Diocese of Massachusetts. This team sponsors programs and provides speakers on environmental topics, including the Creation Tithing Program, for which the Primer is intended as a workbook. The goal of the program is to stimulate ideas, conversation, and action by recognizing that climate change is the most pressing moral crisis of our age.

Anne and her husband live in Cambridge, Massachusetts. They have three children and five grandchildren. For more information about her books visit: www.AnneIpsen.com.

Ken Kroes is an environmental writer living near Calgary, Canada. His works include "Feasible Planet - A guide to more sustainable living" and the "Percipience" Eco-Fiction Series. His passion about sustainable living is due, in part, to experiences gained through extensive travel and living in many locations. His diverse background, which includes agriculture, engineering, and information systems, comes through in his environmental writing along with his strong faith in God.

He strongly believes that governments and businesses need help from each of us in solving our environmental issues. This Primer is a perfect example of outlining how each of us can do more and he is thrilled to have been part of its development.

Ken has a bachelor's degree in Mechanical Engineering and is an active member of the Okotoks Evangelical Free Church. For more information on his work visit: www.feasibleplanet.com.

Made in the USA
Middletown, DE
14 February 2021

33778296R00029